CAN YOU FIND ME?

A Book About Animal Camouflage

For my friend John

ISBN 0-590-41553-0

12 11 10 9 8 7 6 5 4 3 2 1   4 5 6 7 8 9/9

Printed in the U.S.A.   14

CAN YOU FIND ME?

A Book About Animal Camouflage

JENNIFER DEWEY

SCHOLASTIC INC.

New York Toronto London Auckland Sydney

A tawny lion hunts for food, its sleek coat
the color of dry grass.

lion

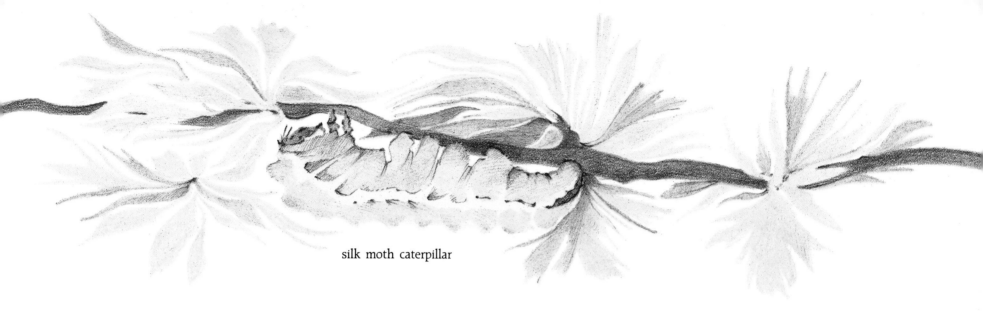

silk moth caterpillar

A striped caterpillar hangs upside down on a twig,
its skin green, shiny, and smooth,
like the leaves around it.

sponge crab

A live sponge grows on the back of a sponge crab.
Covered in this way, the crab is hard to see
as it crawls over the sea bottom, searching for food.

The feathers of a bobwhite match the colors of the forest.
Sitting perfectly still on its nest, protecting
its young, the bobwhite is nearly invisible.

bobwhite

Slithering silently on sandy soil, a desert rattlesnake
travels unseen, its scaly skin the same tans, browns,
grays, and whites as the world it lives in.

These animals, so different from one another,
share a need to hide and be hidden.
They hide to escape an enemy, or to surprise prey.

No animal is safe from its enemies. Danger is everywhere.

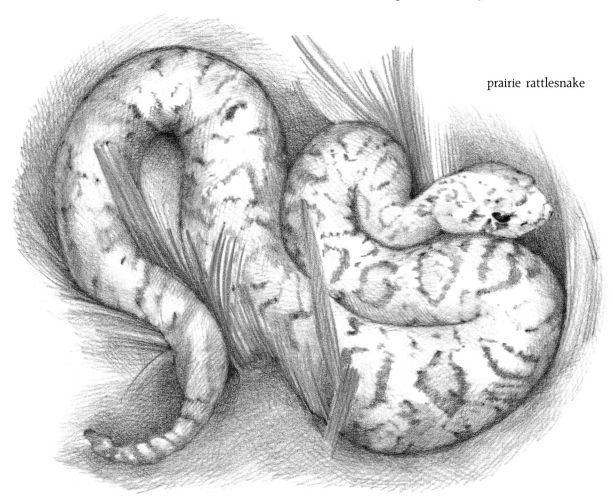

prairie rattlesnake

Animals can fool an enemy by looking, smelling,
or sounding like something they are not.
Butterflies that taste good are colored
like butterflies that taste bitter.
Birds avoid both kinds of butterflies.
Certain snakes leave false trails of scent
to lead an enemy away from their true hiding places.
A dormouse scares its enemies away
by hissing like a snake.

Camouflage is the word for tricks of color and shape
animals have to fool their enemies.
Camouflage is a way to survive.
It helps an animal disappear,
or seem to be something else,
as if it were wearing a costume or a mask.

African grasshopper

crab spider

leaf insects

African monarch butterfly

Camouflage helps an animal find food, avoid being eaten, and live another day.

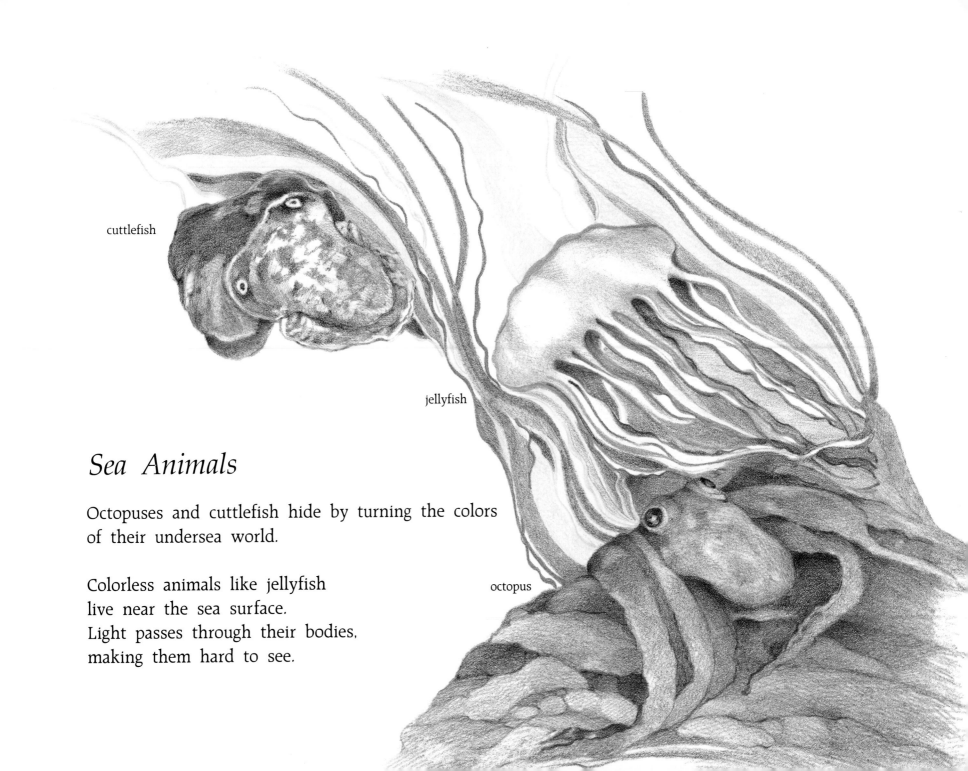

cuttlefish

jellyfish

octopus

Sea Animals

Octopuses and cuttlefish hide by turning the colors
of their undersea world.

Colorless animals like jellyfish
live near the sea surface.
Light passes through their bodies,
making them hard to see.

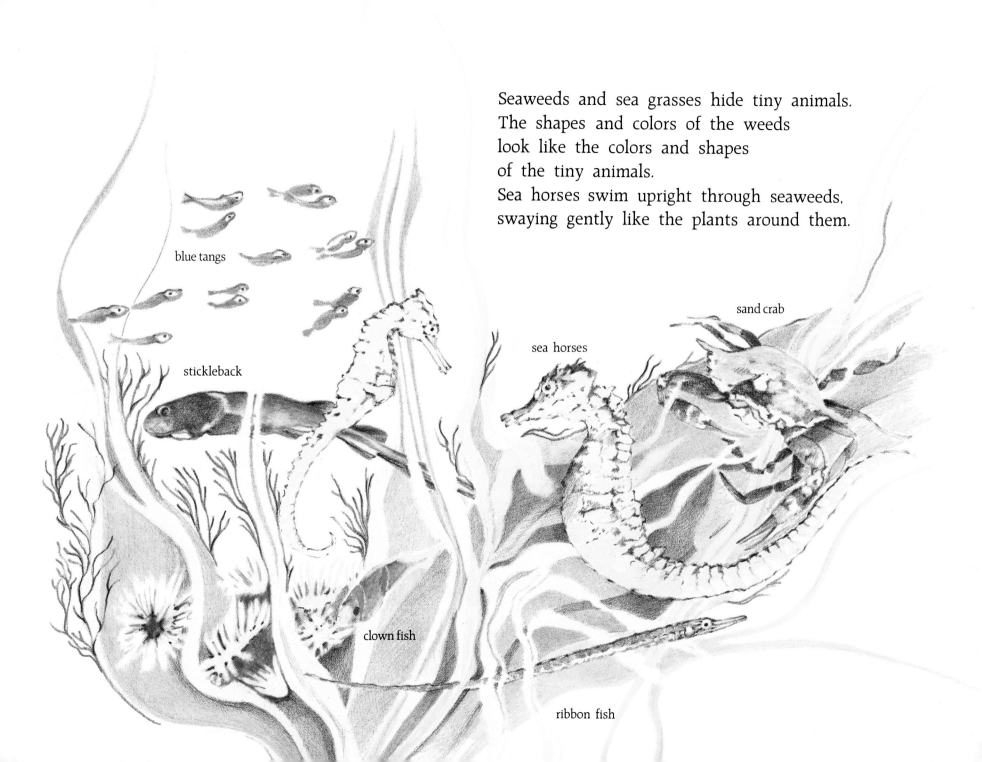

Seaweeds and sea grasses hide tiny animals.
The shapes and colors of the weeds
look like the colors and shapes
of the tiny animals.
Sea horses swim upright through seaweeds,
swaying gently like the plants around them.

blue tangs

stickleback

sea horses

sand crab

clown fish

ribbon fish

eel

butterfly fish

fiddler crab

Eels in coral reefs are brilliantly colored.
Their stripes and spots mimic the coral.

Kelp crabs and common sand crabs slowly
take on the colors of the sea floor
they live on.
Their camouflage evolves over time.

Four-eyed butterfly fish confuse enemies with spots on their tails. The spots look like eyes. Enemies cannot tell if the butterfly fish are coming or going.

A ten-inch-long sea dragon looks more like a floating seaweed than a fish.

sea dragon

eel

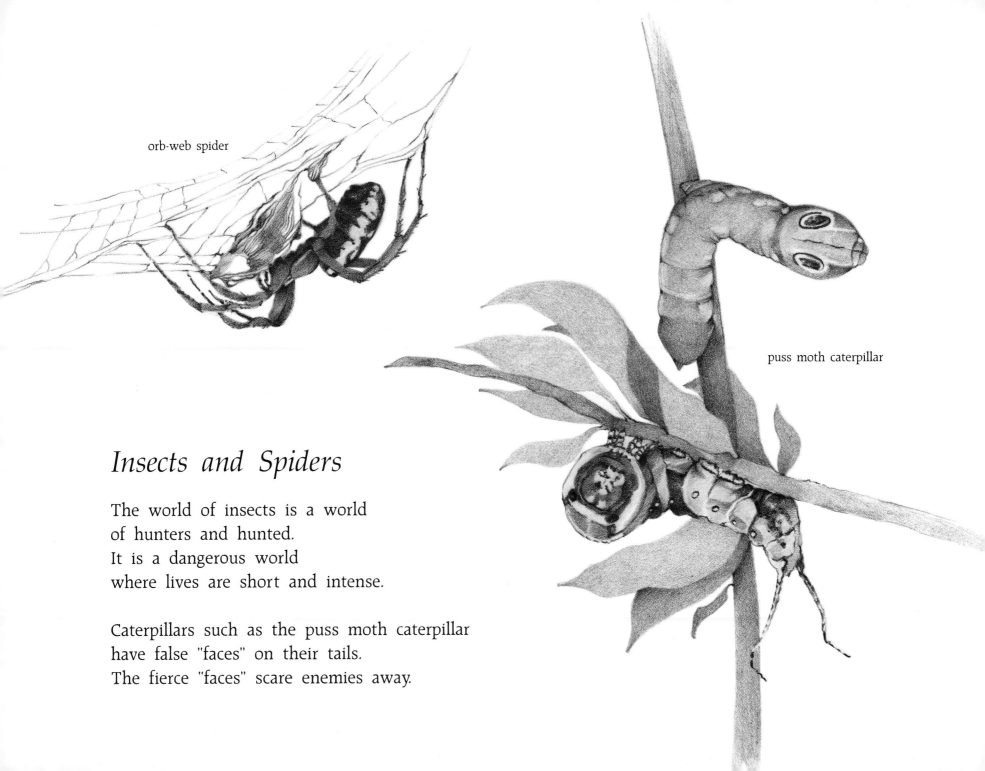

orb-web spider

puss moth caterpillar

Insects and Spiders

The world of insects is a world
of hunters and hunted.
It is a dangerous world
where lives are short and intense.

Caterpillars such as the puss moth caterpillar
have false "faces" on their tails.
The fierce "faces" scare enemies away.

marbled spider

hairstreak butterfly caterpillar

Orb spiders spin beautiful webs. The webs' patterns match those of the spiders themselves.

Hairstreak butterfly caterpillars blend into the cypress trees they feed on.
With their thin, lacy spines and stripes they seem to disappear.

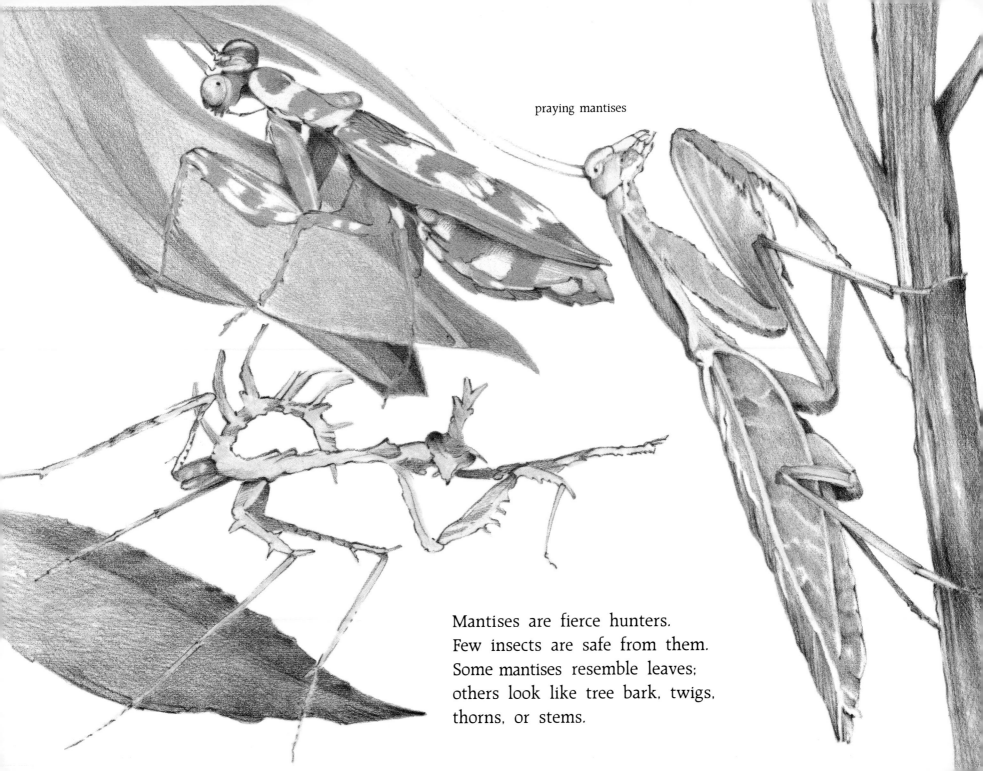

praying mantises

Mantises are fierce hunters.
Few insects are safe from them.
Some mantises resemble leaves;
others look like tree bark, twigs,
thorns, or stems.

tropical flower mantis

The tropical flower mantis
looks like the orchid
it sits on as it waits
to catch insects.
Tricks of shape and color
help to hide mantises
from birds, lizards, and other
animals eager to eat them.

Hiding in a burrow, sheltered by a door
made of silken webbing, dirt, and pebbles,
the trap-door spider pokes out just enough
to capture insects passing by.

trap-door spider

stick insects

There are about two thousand species
of stick insects.
They feed on leaves and flowers.
When an enemy comes near they freeze,
holding so still they look like
the stems and leaves around them.

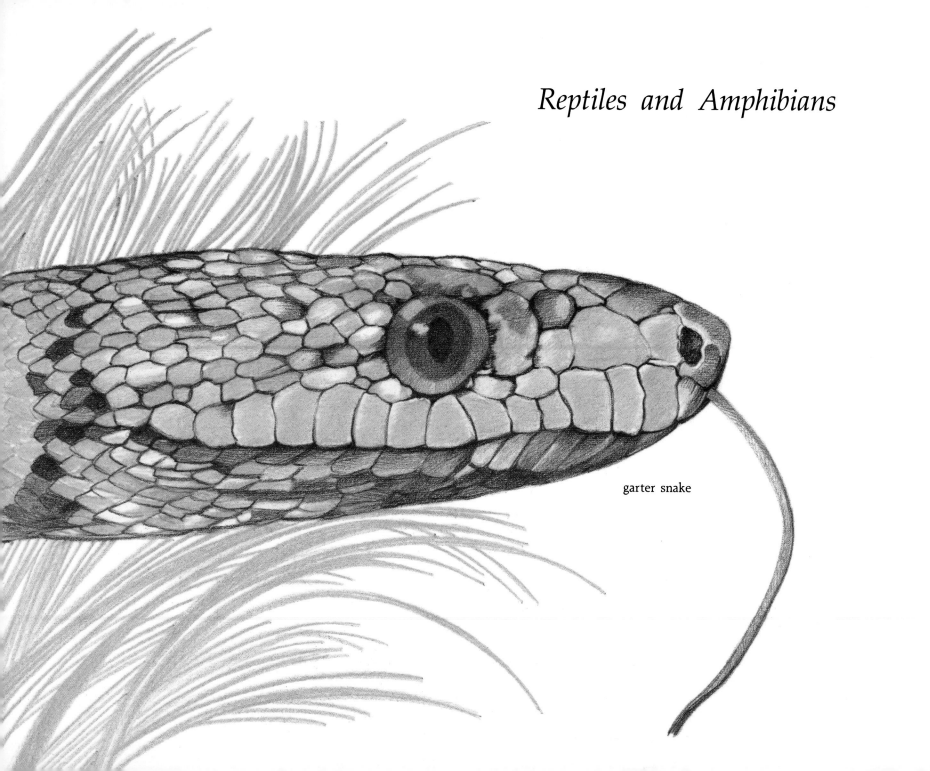

garter snake

Tree frogs lighten or darken because of changes
in temperature, touch, dampness, or light.
They change color quickly,
becoming the shade of the leaf they sit on,
or the stem they cling to.

tree frog

mountain chameleon

Chameleons also react to light, dampness,
touch, and fear.
In the trees a chameleon turns green
to look like leaves. In the grass
a chameleon turns brown or grayish-green.

Asian tokay gecko

The Texas horned lizard is very hard to see.
Its coloring blends with the desert landscape
it lives in.

Texas horned lizard

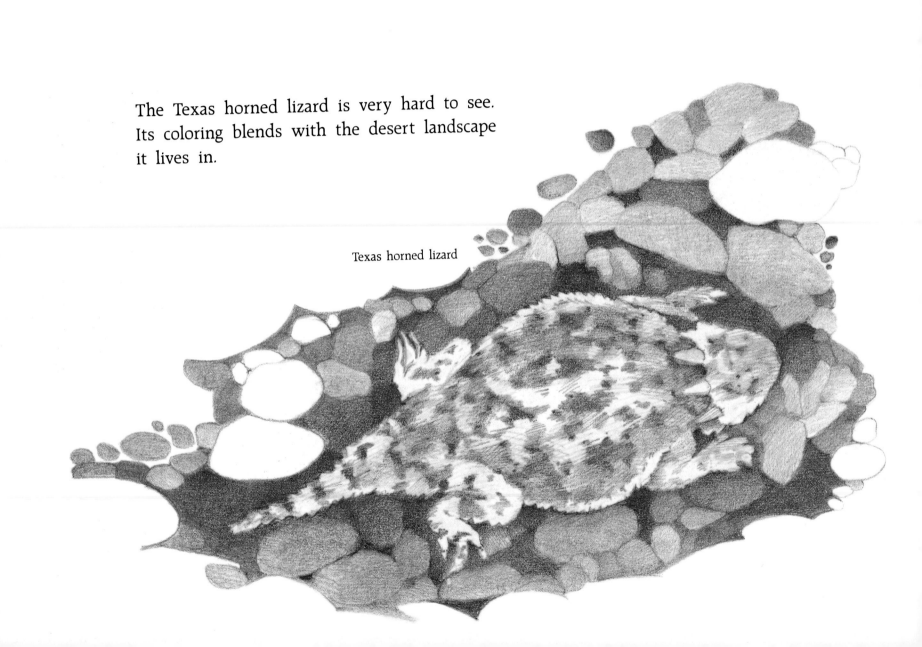

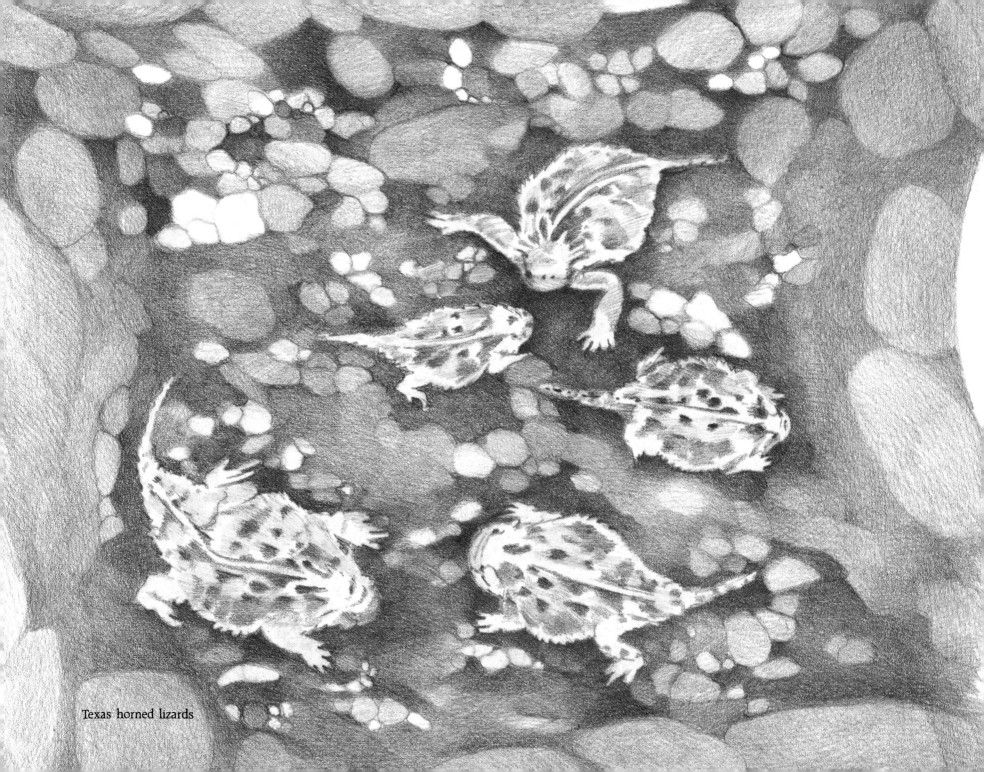

Texas horned lizards

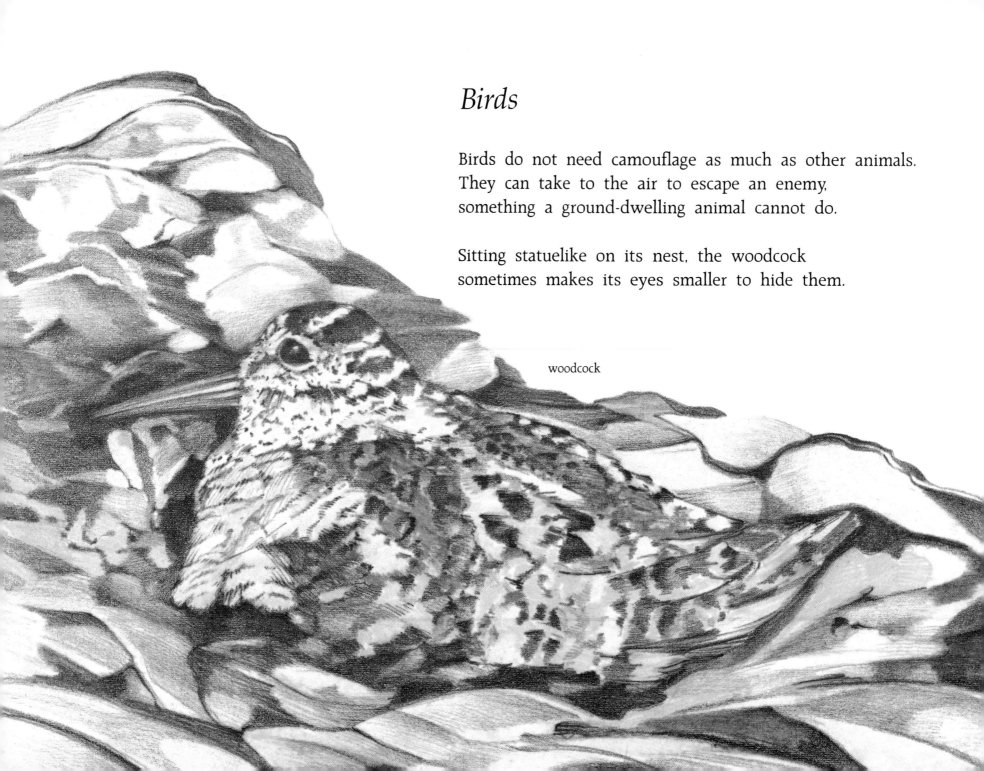

Birds

Birds do not need camouflage as much as other animals. They can take to the air to escape an enemy, something a ground-dwelling animal cannot do.

Sitting statuelike on its nest, the woodcock sometimes makes its eyes smaller to hide them.

woodcock

The potoo's coloring helps it
blend into the tree stump it sits on.
The potoo closes its eyes to slits, staying
perfectly still until all danger has passed.

potoo

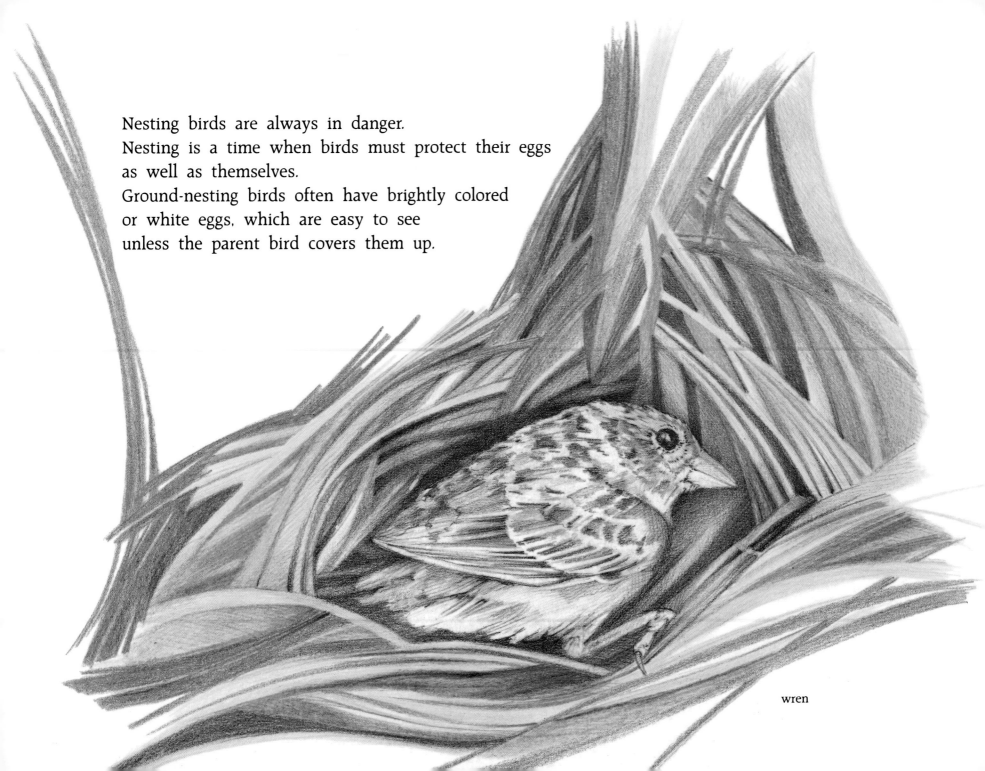

Nesting birds are always in danger.
Nesting is a time when birds must protect their eggs
as well as themselves.
Ground-nesting birds often have brightly colored
or white eggs, which are easy to see
unless the parent bird covers them up.

wren

Bird nests are usually made of materials
that fade perfectly into the background.
The coloring of nests and of parent birds
protects new generations of birds
from enemies.

robin's eggs

For some birds winter means
the world around them turns white.
Snow falls and covers the ground.

Birds living in a wintry white world
must turn white, too, or they would quickly become
a meal for a hungry enemy.

The ptarmigan and snowy owl molt in the fall.
When birds molt, their summer feathers
are replaced with winter feathers,
changing from dark to light.
With new, white feathers
the birds fade into the snow.

ptarmigan

If they are lucky, they will survive until spring.
In the spring they will molt again,
shedding white feathers
for spotted brown feathers that match
their spring and summer world.

snowy owl

Mammals

cheetah

The cheetah, so swift and graceful, with
spots of black and brown against a coat of soft,
yellow-brown fur, can hide in tall grass.

The tiny chipmunk, scurrying up,
down, and around, constantly looking
for food, blends with its
field and forest environment.
Only its bright, shiny, black eyes stand out.

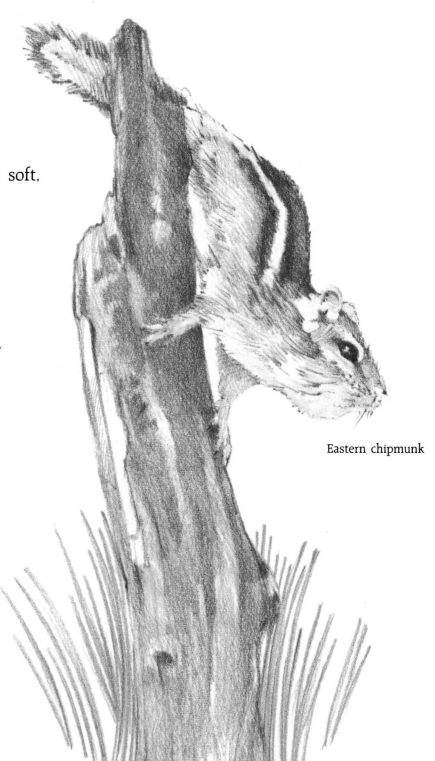

Eastern chipmunk

Mammal babies, like other babies,
must be kept safe from harm.
In nests made of grass, fur, twigs, and leaves,
mice babies tuck close against their mother's side.
They seem almost a part of her
and are safe from enemies.

deer mice

Many mammal babies have
coloring they outgrow when they are older.
This helps camouflage them when they are
very small and not yet able
to protect themselves
from enemies.

The young fawn keeps its spots
until it is almost grown.
The spots begin to fade
only when the fawn
can run very fast and
protect itself from danger.

white-tailed deer

arctic fox

The arctic fox is red and brown in the spring and summer,
changing to pure white, except for its eyes
in the dead of winter.

The snowshoe hare changes, too.
Brown fur is replaced by white.
These animals have a better chance for survival
when they are difficult to see.

snowshoe hare

Animals use camouflage for protection.
Camouflage helps an animal find food,
avoid attack, stay alive, and reproduce its own kind.

People use camouflage.
Some people disguise themselves for protection
against evil spirits.
Others disguise themselves for war, wearing
special clothing to make themselves harder to see.

People also use camouflage when they hunt
so their prey will not see them.

People are aware of using camouflage.
They know they are doing it.
Masks and costumes and feathers and face paint
are deliberate. They are tricks
to fool powerful spirits, or the enemy,
or an animal being hunted.

Singing songs to the sky,
to the mountains,
to the land,
people act out solemn rituals.

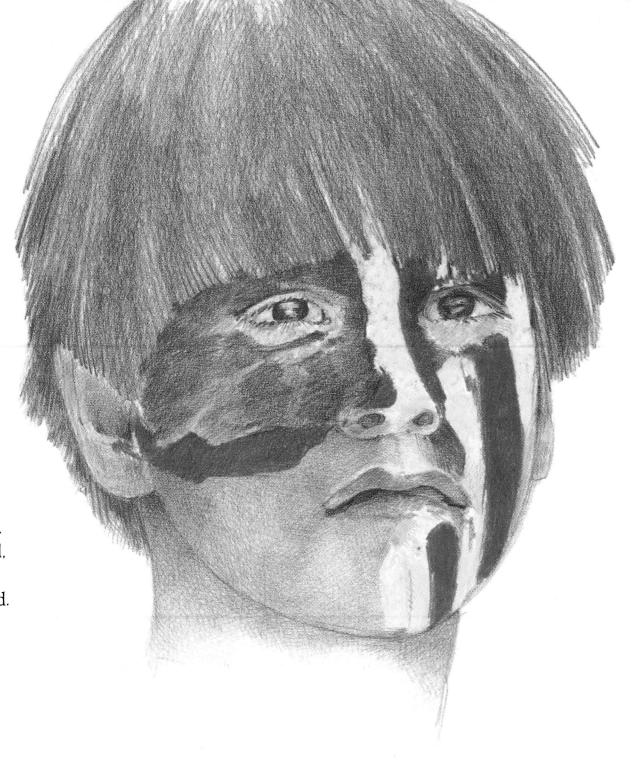

A snake banded in colors
matching the earth
does not know it is disguised.
A Pueblo boy dressed, painted,
preparing for his first dance,
knows and tries to understand.